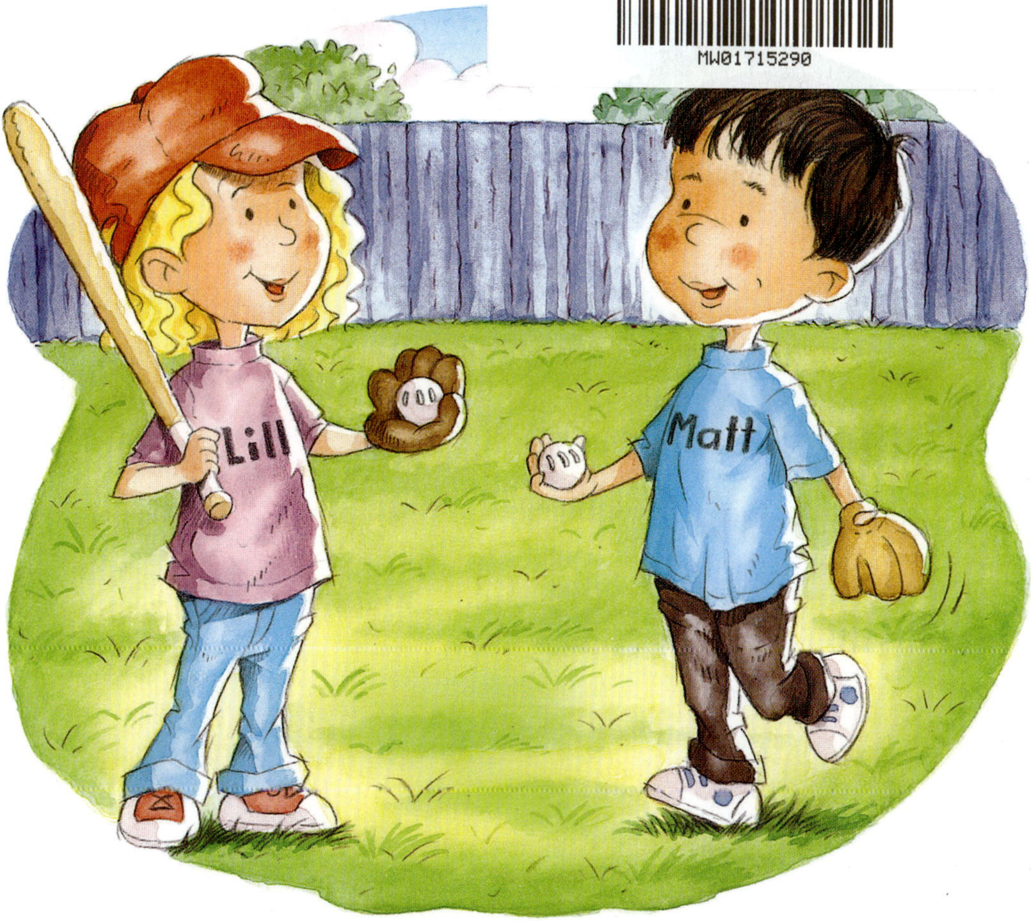

I am Lill.

I am a pal to Matt.

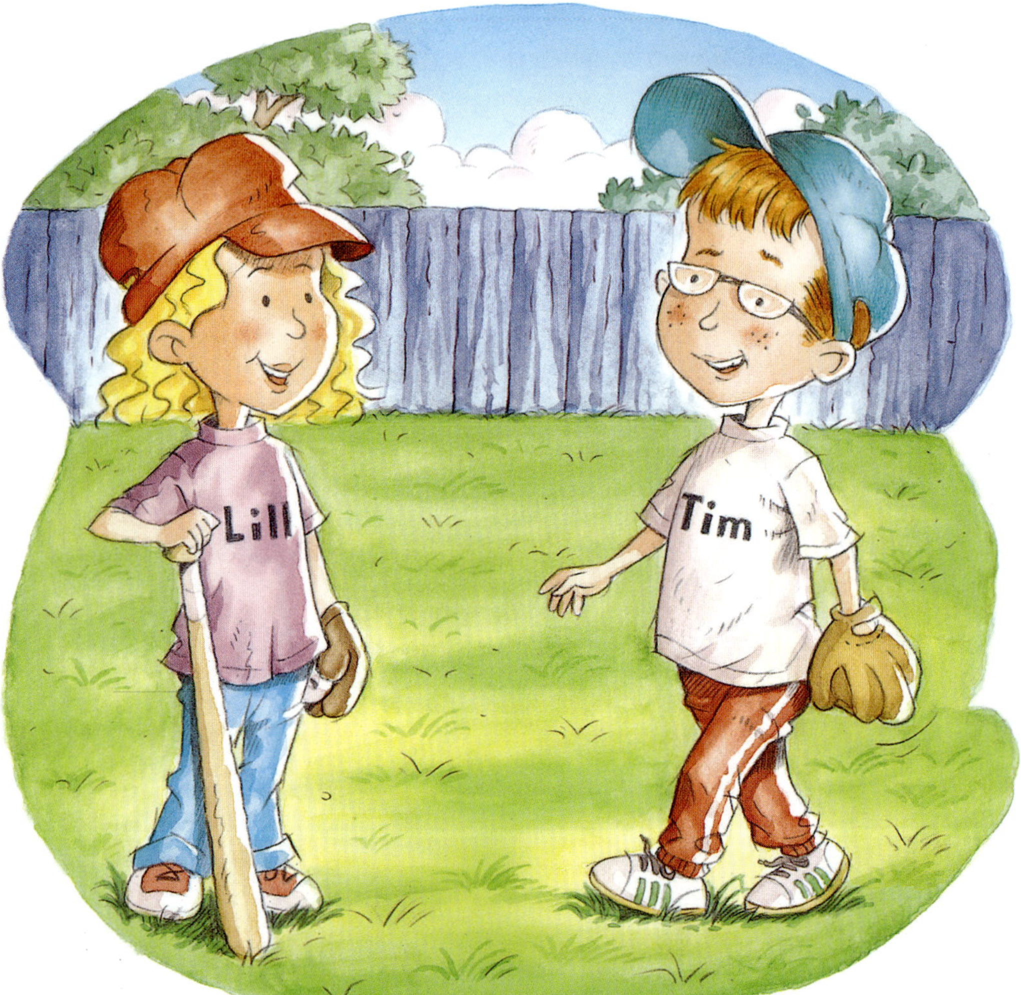

I am a pal to Tim.

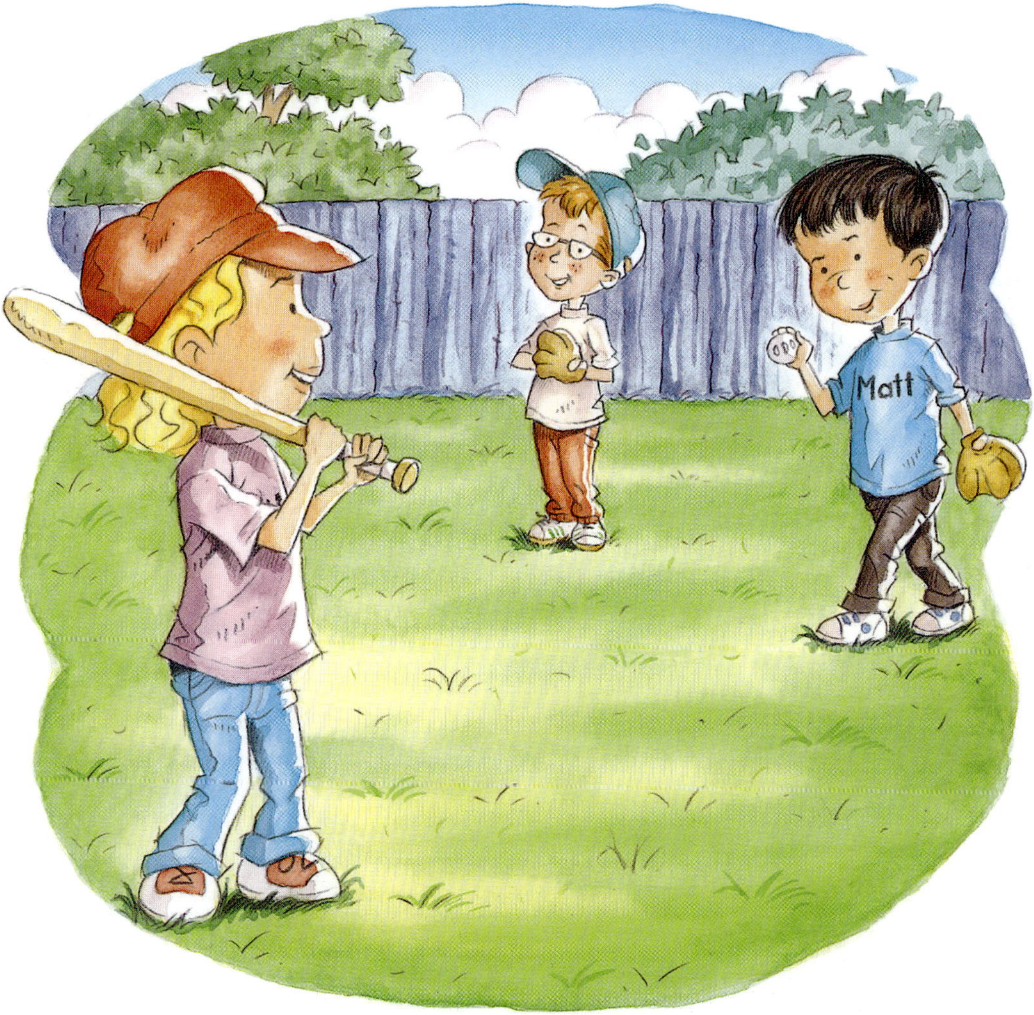

I hit to Matt.

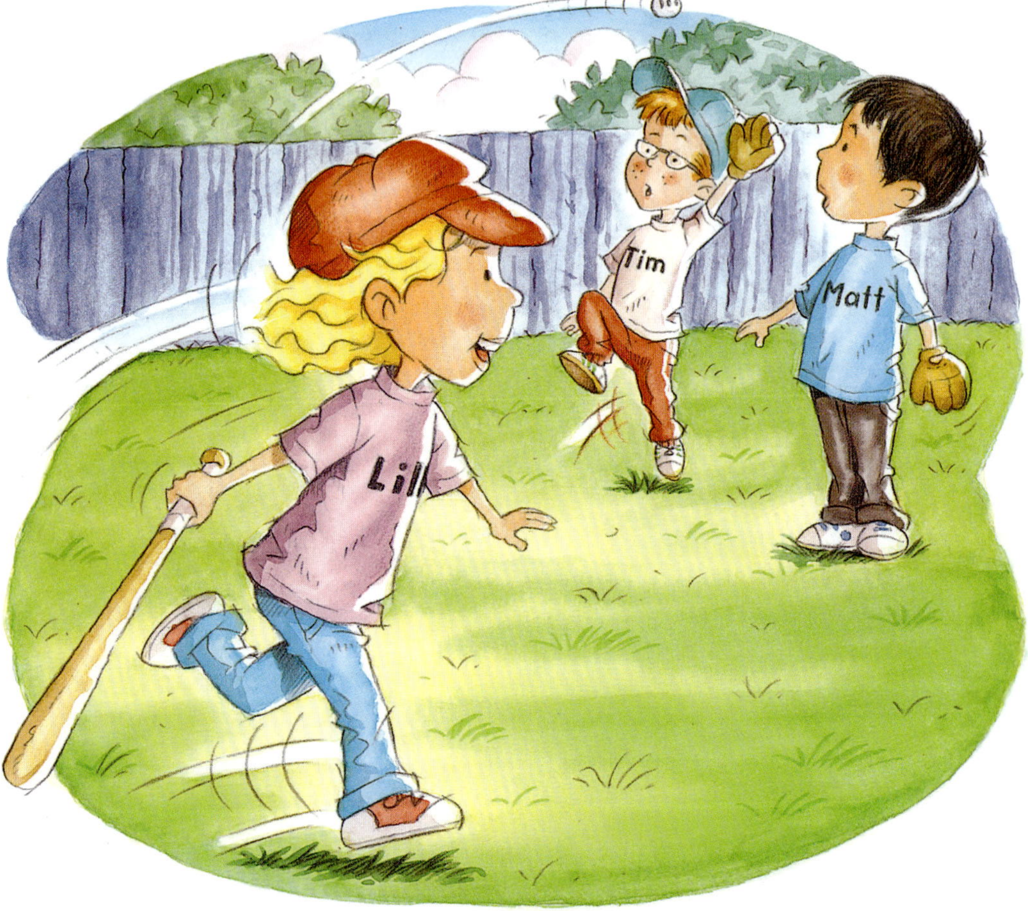

Pop! I hit the 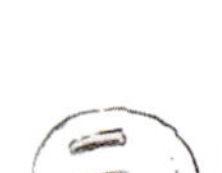 !

I got a hit!

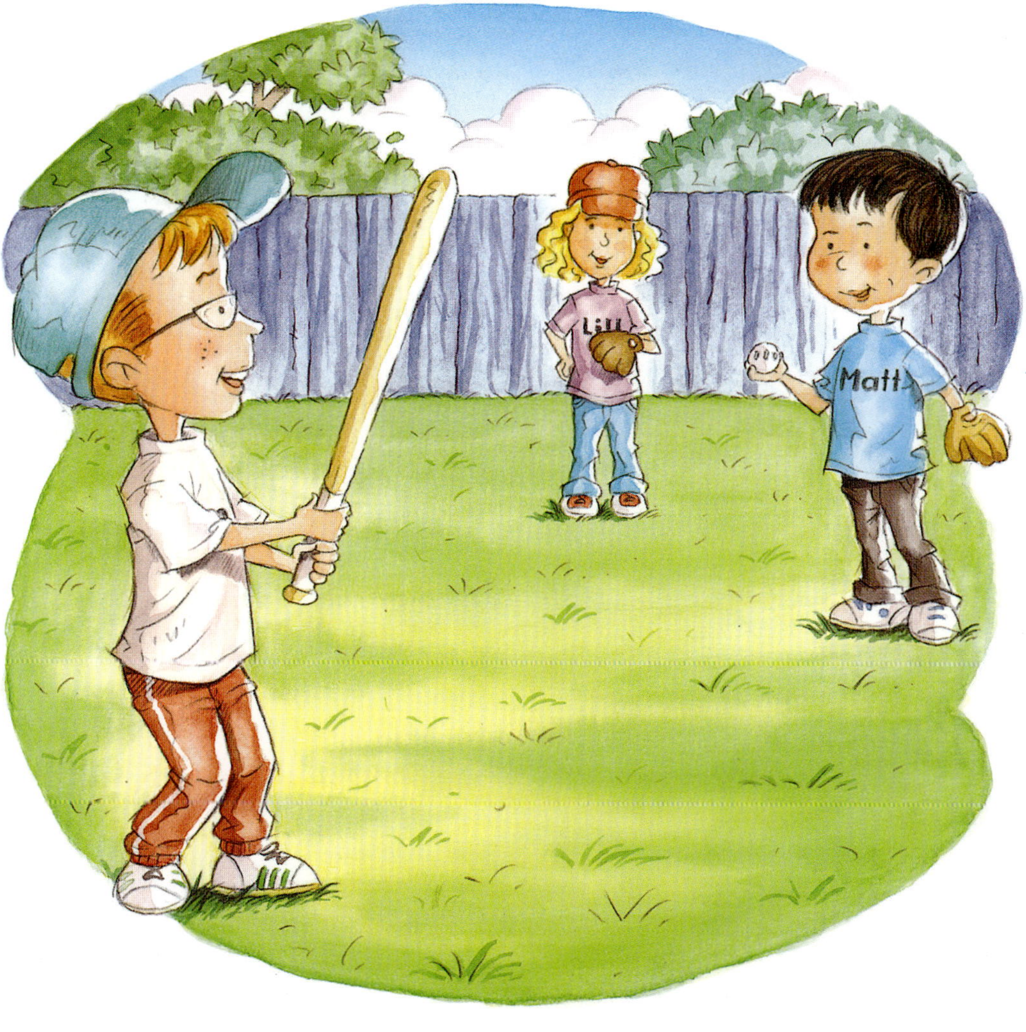

Tim hit to him.

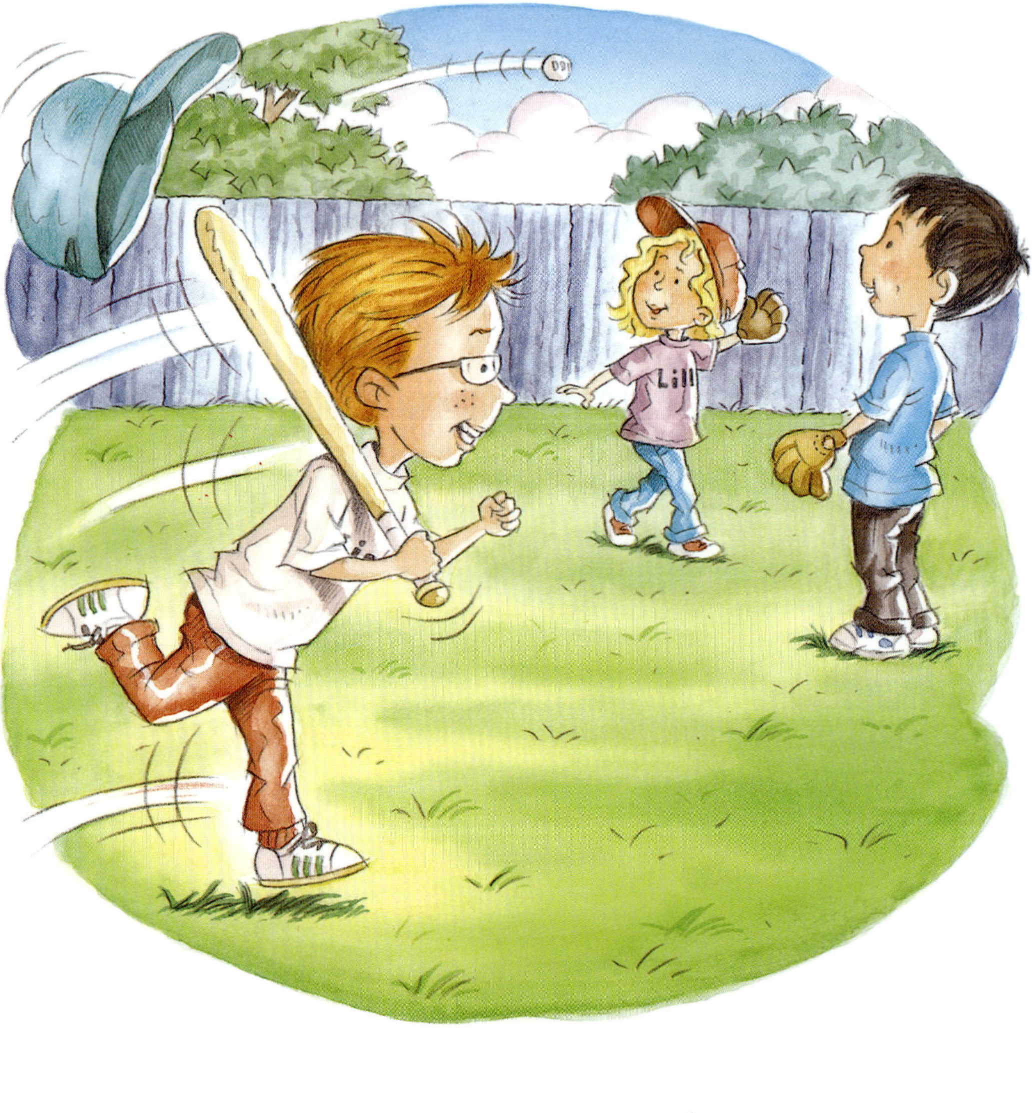

Pop! Tim got a hit?

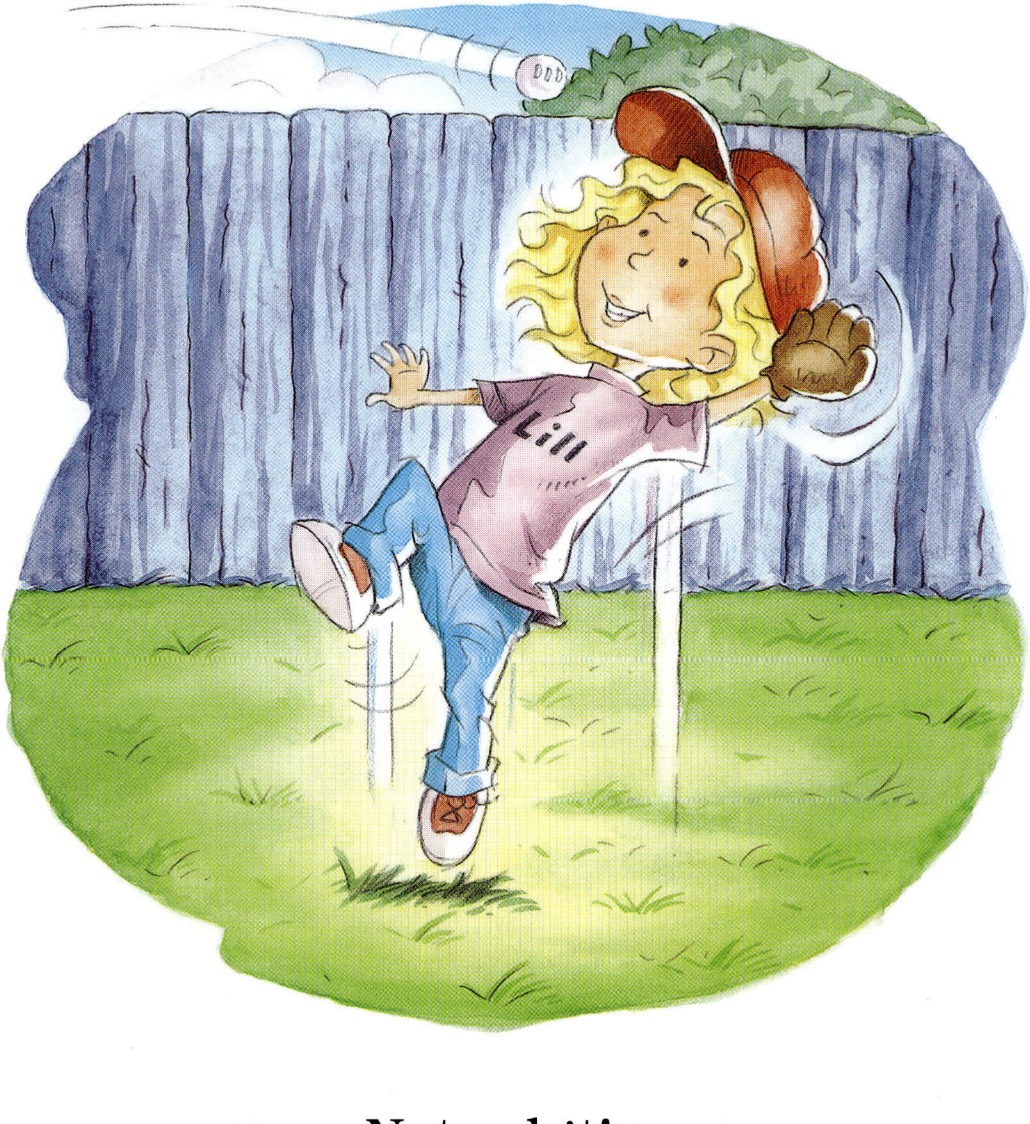

Not a hit!

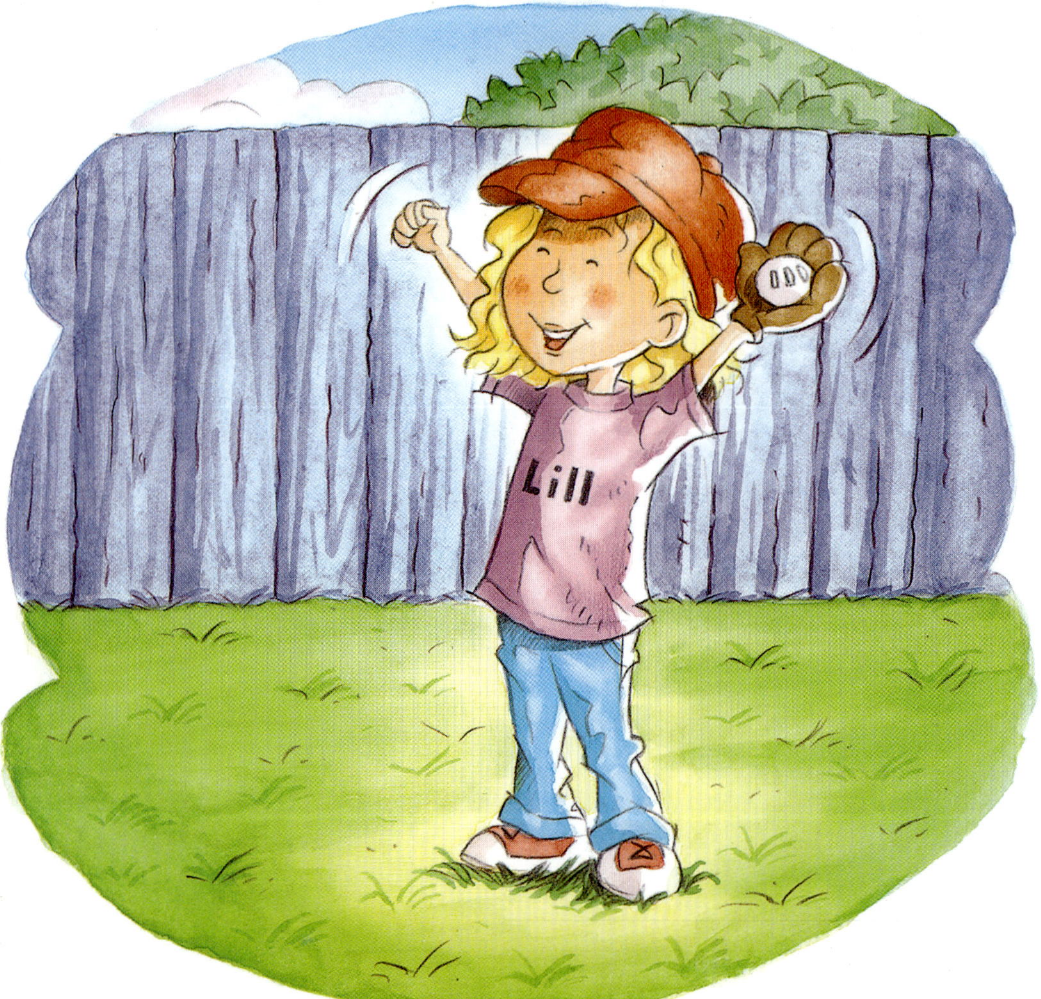

Plop! I got it in the mitt.